NATIONAL GEOGRAPHIC KiDS

美国国家地理
双语阅读

U0179968

Penguins
企鹅

懿海文化 编著

马鸣 译

第三级

外语教学与研究出版社
FOREIGN LANGUAGE TEACHING AND RESEARCH PRESS
北京 BEIJING

京权图字：01-2021-5130

图书在版编目 (CIP) 数据

企鹅：英文、汉文 / 懿海文化编著；马鸣译. —— 北京：外语教学与研究出版社，2021.11（2023.8 重印）
（美国国家地理双语阅读. 第三级）
书名原文：Penguins
ISBN 978-7-5213-3147-9

Ⅰ. 企… Ⅱ. ①懿… ②马… Ⅲ. ①企鹅目－少儿读物－英、汉 Ⅳ. ①Q959.7-49

中国版本图书馆 CIP 数据核字 (2021) 第 236731 号

出版人　王　芳
策划编辑　许海峰　刘秀玲　姚　璐
责任编辑　姚　璐
责任校对　华　蕾
装帧设计　许　岚
出版发行　外语教学与研究出版社
社　　址　北京市西三环北路 19 号（100089）
网　　址　https://www.fltrp.com
印　　刷　天津海顺印业包装有限公司
开　　本　650×980　1/16
印　　张　37.5
版　　次　2022 年 3 月第 1 版　2023 年 8 月第 4 次印刷
书　　号　ISBN 978-7-5213-3147-9
定　　价　188.00 元（全 15 册）

如有图书采购需求，图书内容或印刷装订等问题，侵权、盗版书籍等线索，请拨打以下电话或关注官方服务号：
客服电话：400 898 7008
官方服务号：微信搜索并关注公众号"外研社官方服务号"
外研社购书网址：https://fltrp.tmall.com

物料号：331470001

Table of Contents

What Are They?...........................4

Where Are They? 6

Not Just Any Bird 8

What's for Dinner? 12

Life on Land 16

A Chick Is Born 18

The Longest March 22

Penguin Parade 26

Penguin Play 30

Glossary 32

参考译文 33

What Are They?

EMPEROR PENGUINS

What birds cannot fly?
What birds spend most of their
lives in the ocean but are not fish?
What birds live in the coldest part
of the world—all year long?

They swim, they march, they
slide through the snow.

They are penguins.

Where Are They?

All penguins live between the Equator and the South Pole. Some live where it's very cold. Some live in warmer places like the coasts of Africa or Oceania.

Penguins live on islands, on coasts, and even on icebergs in the sea. They just need to be near water, because they spend most of their lives IN the water.

Not Just Any Bird

EMPEROR PENGUIN

Big webbed feet for better steering.

Layers of soft feathers trap heat. Stiff, oily feathers on top keep out water.

Penguins are perfect for their lives at sea. They have a sleek shape for speed. A layer of blubber keeps them warm.

Stiff flippers act like boat paddles to push and steer.

Big eyes to see underwater.

BIRD WORD

WEBBED: Connected by skin

9

Their black backs make them hard to see from above. Their light bellies make them hard to see from below. But it's their strong, solid flippers that help them escape predators and get where they want to go.

Penguins can swim about 15 miles an hour. When they want to go faster, they leap out of the water as they swim. It's called porpoising (por-puh-sing), because it's what porpoises do.

GENTOO PENGUINS

A predator is an animal that kills and eats other animals.

11

What's for Dinner?

HUMBOLT PENGUIN

Life in the ocean is fish-elicious! Penguins eat a lot of fish. They have a hooked bill, or beak, to help them grab their dinner. Barbs on their tongues and in their throats help them to hold on to slippery food.

Would you like a drink of salty water to go with that fish? Penguins are able to clean the salt out of ocean water. They get fresh water to drink and the salt dribbles back into the ocean.

BIRD WORD

BARB: Something sharp and pointy like a hook

BIRD WORD

MARINE MAMMALS: They have fur and give birth to live young; unlike other mammals, they spend most of their time in the ocean.

While penguins are slurping down their dinners, they have to be careful not to end up as dinner themselves. Penguins are the favorite food of marine mammals such as leopard seals and killer whales.

GENTOO PENGUINS AND A SKUA

Penguins are also in danger on land. Birds like the skua, the Australian sea eagle, and the giant petrel eat penguins. Even cats, snakes, foxes, and rats eat penguins when they can.

Life on Land

KING PENGUINS

BIRD WORD

COLONY: A group of animals who live together

On land, most penguins live in a large colony with thousands or even millions of other penguins. If it's cold, they huddle together. It's so warm inside a huddle that penguins take turns moving to the outside to cool off.

KING PENGUIN HUDDLE

Penguins march together to get to their nesting grounds. Once there they wave, strut, shake, call, nod, dance, and sing to find a mate. Most penguins stay with their mate for many years.

A Chick Is Born

CHINSTRAP CHICKS

Most penguins lay two eggs at a time, but often only one egg survives. The mother and father take turns keeping the egg warm. When it hatches, the parents keep the chick warm and fed.

ADELIE CHICKS

After a couple of weeks, hundreds or even thousands of chicks wait together while the parents go back to the sea to find food. As the chicks wait, they are in constant danger from skuas, eagles, and other animals.

KING CHICK

GENTOO FEEDING CHICK

Finally, the parents return with food. They have to find their chicks in a huge crowd of baby birds. How do they do it? The baby birds sing special songs to help their parents find them.

In a few months, the whole family returns to the sea.

The Longest March

EMPEROR PENGUINS

For the emperor penguins, getting to their nesting grounds is hard work. Their home is Antarctica—the coldest place on Earth.

Emperor penguins nest much farther from the ocean than other penguins. They must march for days and nights through snow and wind.

After laying her egg, the female gives it to the male. He will keep it warm in a flap under his belly. Unlike other penguins, the male emperor cares for the egg by himself while the female goes back to the ocean to find food.

The mother is gone for more than two months. The father huddles with the other male penguins to keep himself, and his egg, safe and warm. During this time, the father eats nothing but snow.

When the mother returns in July, the father quickly goes to the ocean to find food. By December, the whole family is ready to go.

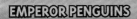

EMPEROR PENGUINS

Penguin Parade

Little Blue
HEIGHT
16"
SMALLEST

Galapagos
HEIGHT
18"–21"

Snares
HEIGHT
21"–25"

Fiordland
HEIGHT
24"

Erect-Crested
HEIGHT
24"–26"

There are 17 different species,
or kinds, of penguins.

Rockhopper

HEIGHT
21"–25"

Yellow-Eyed

HEIGHT
23"–30"

These penguins are the loudest.
They sound like donkeys.

Magellanic

HEIGHT
24"–28"

African

HEIGHT
24"–28"

Macaroni

HEIGHT
21″–26″

Royal

HEIGHT
24″–28″

These penguins are the fastest swimmers.

Chinstrap

HEIGHT
27″

Gentoo

HEIGHT
27″–30″

Adélie

HEIGHT
22″–26″

Humboldt

HEIGHT
22″–26″

These penguins don't make a nest.
They carry their eggs wherever they go.

King

HEIGHT
37″

LARGEST

Emperor

HEIGHT
44″

Penguin Play

HOPPING: Rockhoppers can hop five feet high!

ROCKHOPPER PENGUIN

Life isn't always easy for penguins. But at least they look like they're having fun.

SINGING: Adults sing to their mates, and chicks sing for their parents.

MACARONI PENGUIN

SLEDDING: Penguins speed down icy hills on their feet and bellies to get somewhere fast.

KING PENGUIN

SURFING: Penguins surf through the waves. Sometimes they surf right from the water up onto land.

CHINSTRAP PENGUIN

Glossary

BARB: Something sharp and pointy like a hook

COAST: Where land meets the sea

COLONY: A group of animals who live together

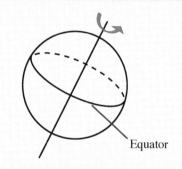

EQUATOR: An imaginary line around Earth halfway between the North and South poles

MARINE MAMMALS: They have fur and give birth to live young; unlike other mammals, they spend most of their time in the ocean.

WEBBED: Connected by skin

▶ 第 4—5 页

它们是谁?

什么鸟儿不会飞?

什么鸟儿大部分时间生活在海洋里,却不是鱼?

什么鸟儿生活在世界上最冷的地方——整年都是?

它们游泳,它们集体行进,它们在雪地上滑行。

它们是企鹅。

帝企鹅

▶ 第 6—7 页

它们在哪儿?

所有的企鹅都生活在赤道与南极中间。一些生活在非常寒冷的地方。一些生活在较为温暖的地方,比如非洲或大洋洲的海岸。

企鹅生活在岛上、海岸上甚至海洋里的冰川上。它们只要在近水的地方就可以,因为它们大部分时间都生活在水里。

北极

北极熊
请这边走

鸟类 小词典

赤道:位于北极和南极正中间、环绕地球的一条假想线

海岸:陆地和海洋交汇的地方

不只是鸟

企鹅非常适合在海洋里生活。它们的身型线条流畅，有助于加速。脂肪层维持它们的体温。

帝企鹅

硬挺的鳍状肢像船桨一样划水，掌控方向。

大眼睛可以观察水下的情况。

大大的蹼足可以更好地控制方向。

层层软毛可以保暖。硬挺的油性羽毛可以防水。

鸟类 小词典

有蹼的：以皮肤相连

黑色的背使它们很难从空中被发现。浅色的腹部使它们很难从水下被发现。它们的鳍状肢强壮又结实，让它们能躲避捕食者，到达它们想去的地方。

企鹅每小时可以游大约15英里（约24.14千米）。如果它们想游得更快，它们会在游水时跃出水面。这种方式叫"跃身击浪"，鼠海豚就是这么做的。

巴布亚企鹅

捕食者是杀死并吃掉其他动物的动物。

▶ 第 12—13 页

晚餐吃什么?

洪堡企鹅

海洋里的生活总是和鱼有关!企鹅要吃很多鱼。它们有钩状的嘴(也叫喙),便于它们收获晚餐。舌头上和喉咙里的倒刺便于它们抓住光滑的食物。

你想在吃鱼时喝盐水吗?企鹅可以把海水中的盐分过滤掉。它们喝淡水,盐分则回海里。

鸟类 小词典

倒刺:像钩子一样锋利、尖锐的东西

▶ 第 14—15 页

当企鹅狼吞虎咽地用餐时,它们还得小心不要被吃掉。企鹅是豹形海豹、虎鲸等海生哺乳动物最喜欢的食物。

企鹅在陆地上也面临很多危险。贼鸥、澳大利亚海鹰以及巨鹱等鸟类都吃企鹅。甚至猫、蛇、狐狸和老鼠有机会时也会吃企鹅。

鸟类 小词典

海生哺乳动物:它们有皮毛,以胎生的方式产下幼崽;与其他哺乳动物不同,它们大部分时间都生活在海洋里。

巴布亚企鹅和贼鸥

▶ 第 16—17 页

陆地生活

在陆地上，大多数企鹅和成千上万甚至上百万只别的企鹅生活在一个大群体。如果天冷了，它们会挤在一起。挤在中间太热了，因此它们会轮流到边上凉快一下。

企鹅集体向它们筑巢的地方行进。一到那里，它们就挥动翅膀，昂首阔步，摇晃身体，高声鸣叫，不停点头，又唱又跳，目的就是寻找配偶。大多数企鹅会和它们的配偶一起生活很多年。

王企鹅

鸟类 小词典
群体：生活在一起的一群动物

挤成一团的王企鹅

▶ 第 18—19 页

小企鹅降生了

大多数企鹅一次产两枚蛋，但是通常只有一枚蛋能活下来。妈妈和爸爸轮流孵蛋，保证蛋是温热的。蛋孵化后，父母要给小企鹅取暖、喂食。

几周后，当父母回到海里觅食时，成百只甚至上千只小企鹅会一起等待。在小企鹅等待时，它们要不断应对来自贼鸥、鹰以及其他动物的威胁。

小帽带企鹅

小阿德利企鹅

▶ 第 20—21 页

终于，父母带着食物归来了。它们要在一大群鸟宝宝中找到自己的孩子。它们怎么找呢？鸟宝宝用独特的歌声帮助父母找到自己。

几个月之后，企鹅一家回到了海里。

小王企鹅

巴布亚企鹅给孩子喂食

▶ 第 22—23 页

最长的征程

帝企鹅

对帝企鹅来说，到达它们的筑巢地的过程极为艰辛。它们的家在南极洲——地球上最冷的地方。

帝企鹅要到远离海洋的地方筑巢，比其他企鹅都要远。它们必须夜以继日、顶风冒雪地行进。

雌企鹅产蛋后将蛋交给雄企鹅。他会把蛋放在腹部的下方。与其他企鹅不同，雄帝企鹅在雌帝企鹅回到海里觅食时独自照顾企鹅蛋。

▶ 第 24—25 页

妈妈已经离开两个多月了，爸爸和别的雄企鹅挤在一起，以保证自己和蛋的安全，也能取暖。在这期间，除了雪，爸爸吃不到别的东西。

当妈妈在七月归来时，爸爸赶紧回到海里觅食。到十二月时，帝企鹅一家准备出发了。

帝企鹅

企鹅大游行

这里有 17 种企鹅。

最小的

小蓝企鹅
身高16英寸
（约40.64厘米）

加岛环企鹅
身高18—21英寸
（约48.72—53.34
厘米）

斯岛黄眉企鹅
身高21—25英寸
（约53.34—63.5
厘米）

黄眉企鹅
身高24英寸
（约60.96厘米）

翘眉企鹅
身高24—26英寸
（约60.96—66.04
厘米）

跳岩企鹅
身高21—25英寸
（约53.34—63.5
厘米）

黄眼企鹅
身高23—30英寸
（约58.42—76.2
厘米）

这种企鹅的叫
声最响亮。它们听
起来像驴子。

非洲企鹅
身高24—28英寸
（约60.96—71.12
厘米）

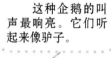

麦哲伦企鹅
身高24—28英寸
（约60.96—71.12
厘米）

▶ 第28—29页

这种企鹅游得最快。

白颊黄眉企鹅
身高24—28英寸
（约60.96—71.12
厘米）

马可罗尼企鹅
身高21—26英寸
（约53.34—66.04
厘米）

帽带企鹅
身高27英寸
（约68.58厘米）

巴布亚企鹅
身高27—30英寸
（约68.58—76.2
厘米）

这种企鹅从不筑巢。它们走
到哪儿，就把企鹅蛋带到哪儿。

最大的

阿德利企鹅
身高22—26英寸
（约55.88—66.04
厘米）

洪堡企鹅
身高22—26英寸
（约55.88—66.04
厘米）

王企鹅
身高37英寸
（约93.98厘米）

帝企鹅
身高44英寸
（约111.76厘米）

▶ 第30—31页

企鹅玩耍

　　对企鹅来说，生活并不总是那么容易，
但至少它们看起来非常开心。

歌唱：成年企鹅用
歌声吸引配偶，小
企鹅用歌声让父母
找到自己。

帽带企鹅

跳岩企鹅

跳跃：跳岩企鹅可以跳5英
尺（约152.4厘米）高！

马可罗尼企鹅

滑雪橇：为了迅速到达某
个地方，企鹅会借助脚和
腹部从冰山上滑下来。

王企鹅

冲浪：企鹅在海里冲浪。
有时它们会从水里一直冲
到陆地上。

39

词汇表

倒刺：像钩子一样锋利、尖锐的东西

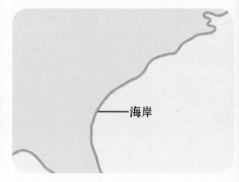

海岸：陆地和海洋交汇的地方

群体：生活在一起的一群动物

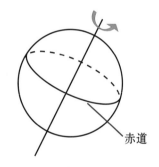

赤道：位于北极和南极正中间、环绕地球的一条假想线

海生哺乳动物：它们有皮毛，以胎生的方式产下幼崽；与其他哺乳动物不同，它们大部分时间都生活在海洋里。

有蹼的：以皮肤相连